The Force That Acted

By: Andrew S Edwards

Introduction

Do you realize that as you are holding this text in your hand, you are reading an entire book that contains nothing but different combinations of 26 letters? Why do you understand it? Is it even real, or is your mind just telling you what you want to believe? Does believing that it is real, therefore make it real? When thinking of the existence of a Godly or superior being, it is somewhat the same. Our minds tell us it is there; this is something our minds have told us over thousands of years, and that belief has not gone away. There is an argument that even the fact that we can think of a Godly being is proof in itself, since every thought is born in some reality. The dream of flying over the earth is born in the fact that we watch birds and wish for their gift of flight. Now, try to picture a color that does not exist, a color that is no combination of any known colors, or anyone has ever seen. You cannot. All thoughts are born in reality. Even still, the existence of a Godly being is questioned because you cannot touch, feel, see, or hear it- yet can we even prove to ourselves that our decided reality is real?

I hope this first paragraph has you thinking, as I hope you continue to do as you read through this. The only thing I ask is that anyone reading this keeps an open mind. Question everything, but be willing to accept it as well. The first part of this writing is about the existence of a God. It takes no stance as far as religion, but opens your mind to the existence of God. The second part of the writing introduces religious concepts of the Abrahamic faiths, though I have tried to keep it open to many interpretations including Judaism,

Christianity, and Islam. Try and think of parts one and two separately, as if they are two separate pieces of writing.

It has taken me a several years to compile all of these thoughts, and many still remain questionable to me. Years of research, reading and watching information on religion and atheism, and questioning everything I know. I have felt for a long time that too many ideas in this world go one way or the other stating either "there is no God" or "our God and religion are the right God and religion." Not that they are all that cut and dry, but very few leave much up to personal interpretation so that someone can follow what their heart and mind are truly telling them is real. It also seems many are unwilling to accept religion as part of their culture because of negative feelings in their personal past, though in many cases it is their own culture and their family's culture that they are giving up. I hope this writing allows that feeling for those that read it, as well as opening the door for new concepts to the reader.

With all that being said, please do not dismiss or accept any concept until you have read it all. Perhaps until you have read it several times. Everything written here goes together, and by judging one piece too soon, you may lose another piece without even thinking it through. Therefore, without any more delay, I will begin.

Part 1

"I find it as difficult to understand a scientist who wouldn't recognize the existence of a superior rationality behind the existence of the universe, as it is to comprehend a theologian who would deny the advances of science." -Warner Von Braun

Let me start by explaining that this part of my theories and writing is about a superior being in general (i.e. - God, Yahweh, or Allah) and not about religion. Religion and belief in God are two different things. Religion is the history and practice of reaching God and/or heaven, while God in general is a being or force. In my eyes, all prayers go to one God, no matter the religion, because there is one God to hear them. If any person believes in monotheism, they therefore must believe that there is only one recipient of prayers and he/she hears them all, no matter his/her name. I will not state my own religious beliefs because, once again, this is not about religion.

     I will attempt in this theory to show there must be a superior being working on the universe, a sort of supernatural force. At the heart of this all is the laws of physics, the laws of nature, and general mathematics each with their own attributes. In the end, for those of you still unable to accept that an unknown force exists everywhere, yet is untouchable and unseeable, I will use science to show how, in reality, zero and infinity are based upon the same principles and could even be the same.

     As we begin this theory, I again urge you to read it without preconceived notions whether for or against, and try to understand the theories rather than attempting to support or disprove them in your mind. Whether you agree with the theories or not can be determined after you are done. Either way, I hope you enjoy the ideology behind all of this, and can perhaps learn something about physics and math, if not God. I additionally hope that everyone reading this can see the comedy in the fact that the one science banned by early church officials and religious scholars is now the science attempting to prove their beliefs.

***

Physics:

I am sure most of you reading this have heard of Sir Isaac Newton and his laws of gravity and inertia, and if not, you soon will. Sir Isaac Newton stated that an object in motion will stay in motion and an object at rest will stay at rest until acted upon by another force. Think of it as a billiard ball- it stays in one place until something hits it. Of course, here on earth an object in motion will not stay in motion because of gravity and wind resistance, but you can see that a rock will not move unless it is pushed or gravity pulls it when dropped.

Inertia is a continuation of that law, in that if an object is moving in one direction and is forced in another direction, it will move in both directions until one force is more powerful. To better understand this think of riding in a car, when you turn fast you start to lean the other direction and your coffee spills the opposite direction from the one you are turning. That is because the forward inertia of your body is forced in another direction and the forces of motion must equal out. In this sense, you are not directly turning right at 90 degrees, you are turning right at different angles until you are at 90 degrees in a curved turn rather than a direct change of direction.

With this understanding, we now look at space- a vast emptiness of rapidly expanding emptiness containing continuously orbiting balls of rocks and gas and vast fusing particles. It is the properties of space, along with the laws of motion and inertia, which initially prove the work of a superior being starting with the big bang theory. The big bang theory is, of course, the theory that everything in the universe was compacted into one infinitely dense speck of matter that exploded and expanded into the particles and universe we have today ultimately causing cosmic background

The Big Bang and Universal Expansion.

All matter and energy in on infinate point... at rest.

radiation. Basically, the universe was packed so tight, it was one little dot until it exploded outward. The question we must ask is, what created that infinitely dense speck, and what caused it to explode into the perfection of life we have now? According to Sir Isaac Newton's laws, an outside force must have acted upon it. An outside superior force that either made this point, set it into motion, or both.

Even further within the big bang theory is that this infinitely dense speck of matter would have had an infinite gravitational pull. Since mass is the property that defines gravity, this speck of matter that occurred just before the universe exploded and expanded would have more gravitational pull that all the black holes combined- which not even light can escape. Therefore, an outside force capable of not only acting upon this object at rest, but also causing it to break its infinite gravitational pull would be quite a feat, which no known physical property today could break other than a superior being.

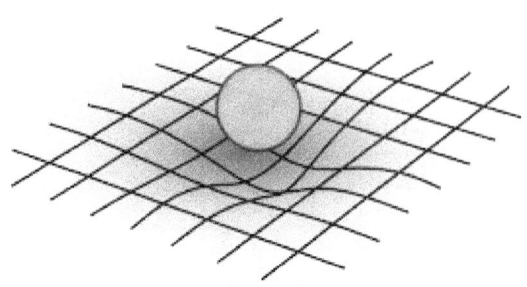

Gravity Causes Space to Bend Around Mass

Let us fast-forward a moment though. Today, we have space ever expanding, always in motion, and the earth in perfect orbit of the sun. To understand this orbit you must understand Einstein's general theory of relativity. This theory states that anything with mass must have gravitational pull based on that mass. Picturing this within space, is like when you put a rock on an open blanket and it causes the blanket to cone around it. Then by placing a marble on that blanket, it will move inward to the rock. Einstein asserts that structure of space is the same way, pulling matter towards objects of mass. This knowledge raises so many questions about how so many of the natural phenomena occur in space.

How was the earth not been pulled all the way into the

sun at a 90-degree angle when even orbiting satellites eventually fall back to earth, especially when the rocks and gasses must have collided together from somewhere else? What force acted upon the earth to make it spin in addition to its movement around the sun? It seems like some outside force must have been pivotal in causing all the motion in space. Everything is beginning motion, everything's change of direction, motions placed perfectly in play and set off by some unknown force that in the end resulted in our existence. Even those that believe in coincidence must believe the laws of nature and that something set it all off.

To complicate the matter more is everything's motion through the fourth dimension of space- time itself. As Einstein had suggested, time is additionally a dimension on the grid of x, y, z (the dimensions we see in, since we see everything in 3d) so surely the laws of physics also apply to the motion of time, which must have been acted on my something, but that it is a more theoretical area.

The final laws of physics I will bring up are the laws of conservation of mass and the laws of conservation of energy written by Antoine Lavoisier. These laws state that you cannot destroy mass, because the matter will break and create something of the same mass or several things that equal out to the same mass. The same goes for energy. Even when you run a car the energy from the gasoline become heat energy, kinetic energy (motion), and expels and exhaust that has some energy itself. With that, you must look at a dying human and say, what will become of that energy? What will become of the electricity of the nerves and the energy that lets us walk and move? What will happen to their conscience? Surely, one cannot believe in the destruction of conscience and mind, for it itself has the ability and strength of any energy. Just as a mothers sense of love can cause chemical reactions to give strength enough to save her child, the human mind is a source of unknown energy that therefore cannot be destroyed.

Further yet, with the two final laws, we must look again to the big bang theory. Where did that infinitely dense point of mass and energy come from? If it cannot be destroyed as told by the laws of conservation of mass and energy, how then was it created? How is mass and energy created if it cannot be destroyed? The only possible source according to the laws of physics must be a superior almost supernatural force.

***

Properties of Zero and Infinity:

At this point the argument over this superior supernatural force would be that it is everywhere but cannot be seen, felt, or touched- everywhere and nowhere at the same time. It has both the properties of zero and infinity. To the human mind this is impossible, outrageous, and is just as hard to believe as an infinitely dense point of mass and energy exploding into the universe. In reality though, zero and infinity have many of the same properties and could even be considered as one in the same based off of numerous examples.

We again turn to movement to prove this point. If you place a speck of something into space with no forces acting on it and with absolutely no movement, it still moves in two different ways though its movement and force are both zero. First, it moves through space, because space is infinitely expanding. That speck of something will constantly be further from the center of the universe as well as further from the edge of the universe though it is making no movement. Secondly it will be in constant movement through the forth dimension on the continuum grid, which is time (the other dimensions being x, y, and z which means we see in three dimensions or 3D with time represented as "t", and not flat 2D), in which all things are constantly moving. Therefore, in two parts of the space-time continuum, or grid, this tiny bit of anything has both zero and infinite movement.

The knowledge of space alone is enough to prove the coexistence and co properties of zero and infinity. For the very fact that it is infinitely expanding but at the same time is absolutely nothing, neither heat nor particle, is mind blowing in itself. Space itself has both the properties of zero and infinity.

Perhaps though, we should look at more worldly examples of the properties of zero and infinity. For this, we go to inflation. Inflation is the statement that the more of something you have, the less value it has to you. For instance, if you had one billion grains of salt, one grain of salt would have probably zero value to you. The same property applies to the value of infinity. Therefore, if it were possible to have an infinite number of something, it would have zero value to it, like the air we breath everyday, but don't think twice about.

The final bit of evidence for the properties of zero and infinity is in mathematics, multiplication and division to be exact. Take any number in the world and multiply it by zero and the result will always be zero,

$$\infty \times 0 = ?$$

no matter what. Now multiply any number in the world by infinity and the result will be infinity, no matter what. That very same principle applies to division, as it is not possible to divide a number by either infinity or zero without infinity or zero being the result. Both infinity and zero have the same properties in multiplication and division.

Time after time, example after example, we see that zero and infinity only have the same properties, but overlapping properties. This is undisputable evidence that it is more than possible for something to be infinite yet at the same time not touchable, seeable, hearable, or smellable… hence zero.

***

Psychology:

Even here, I know the strongest willed Atheists will bring forward arguments of "why wont God (or a superior being) allow us to see it?" Why, if there is a God, would it allow suffering? And more arguments of the same nature, but this is not the place to answer those questions, as those are questions of religious nature. Therefore science alone cannot answer those questions, and I challenge you to find them for yourself if you so choose, for perhaps we do see this superior being in everything around us. Perhaps it is just so superior and great our minds cannot fathom seeing it. Perhaps this superior being does not want to interfere overtly in our human endeavors so we may learn a lesson ourselves, or perhaps it does interfere and we just choose not to see it. This all leads to the psychological questions of a superior being, as our minds seem to be formatted to know and understand this superior being and to seek it as an undying thirst. This is how psychology proves the existence of a superior being- whether through every day actions, or through the actions of the masses throughout history.

I recently had a conversation with someone on the topic of pain. Pain, suffering, and sadness to be exact. It hit me in this conversation that without sadness, without pain, and without low points in our lives we would not know happiness, glee, or any feelings of exhilaration in our lives or come to live for these moments because all moments would be painless and simple. Think of a sad painting- a picture of war that draws your sympathy. This emotion of sadness is necessary to us, and without it we would be nothing. This leads to a worldly thought of how the mind works- how pain can lead to happiness- how happiness and different ideas of the mind can lead to show the hidden aspects of a world designed by a greater supernatural being.

Other than the fact that, as human beings, we are afraid of the dark unknown aspects of the world, how does this strange existence of mental thought affect us? The world is our answer. The colors of the world in particular. When looking from afar, or near, there are three colors of the world that are prevalent: green, blue, and yellow. Life and plants, ocean and sky, and finally desert and sun respectively. These three colors dominate our lives and our earth for a reason- a reason that cannot be coincidence, for the psychological effects of them alone are profound. From a psychological point of view, every color has a meaning and an effect on the emotions of humans, particularly those most common to us.

We will start with green. The color green, often used in hospitals, produces chemicals in the human brain that calm people down. It is thought to have the perfect balance of cool and warm colors that causes a sensation of tranquility and refreshing relaxation. Next is blue, the color of both the sky and the waters. Blue also creates chemicals in the human mind that causes calming tranquil effects. Companies such as Wal-Mart and Facebook use the effects of the color blue to relax you and make you want to stay longer. Blue is often thought of the opposite of red-, which causes aggression. Finally is yellow, the color of the sand and sun. This color is symbolic of happiness, joy, and good times. All of these colors have been proven to affect the human mind, yet all cause a calm happiness in us. Perhaps these colors affect us because they are so prevalent, or perhaps they are so prevalent because they affect us so well, but this is the same chicken or the egg problem. I feel it cannot be coincidence that the peace and joy that we all seek is found everywhere on earth, in the three most common colors we see. I would say it must be superior design that the earth around us causes these effects. When one asks, "Why is the sky blue?" it is not a mere matter of reflection alone, but a matter of design in science.

The final argument in the matter of psychology in the presence of a superior being that I have to bring forward comes straight from the court of law. In the court of law, the strongest piece of evidence one can bring forward is witnesses. The witnesses in this instance are the countless majorities generation after generation, century after century spanning every race and continent who believed in some sort of supernatural being or force as if something in their heart and mind led them to naturally believe. It is the stories of prayers and miracles brought forward by these people that speak volumes in this argument. It is the chilling stories of those that have been dead for mere minutes only to be brought back by doctors, who tell of their immediate conversion due to the horrors they saw in death. It is these countless numbers that forever held the majority on this earth that I feel is the strongest piece of evidence in this story.

In the realm of psychology and the ideas of good and evil, the final piece of information I will leave you with in this portion of this writing. It is a short story that has been floating around the internet for some time now. It has been attributed, unconfirmed, to Albert Einstein, but what is important is its message:

The professor of a university challenged his students with this question. "Did God create everything that exists?" A student answered bravely, "Yes, he did".

The professor then asked, "If God created everything, then he created evil. Since evil exists (as noticed by our own actions), so God is evil. The student couldn't respond to that statement causing the professor to conclude that he had "proved" that "belief in God" was a fairy tale, and therefore worthless.

Another student raised his hand and asked the professor, "May I pose a question?" "Of course" answered the professor.

The young student stood up and asked: "Professor does cold exist?"

The professor answered, "What kind of question is that? ...Of course the cold exists... haven't you ever been cold?"

The young student answered, "In fact sir, Cold does not exist. According to the laws of Physics, what we consider cold, in fact is the absence of heat. Anything is able to be studied as long as it transmits energy (heat). Absolute Zero is the total absence of heat, but cold does not exist. What we have done is create a term to describe how we feel if we don't have body heat or we are not hot."

"And, does Dark exist?" he continued. The professor answered "Of course". This time the student responded, "Again you're wrong, Sir. Darkness does not exist either. Darkness is in fact simply the absence of light. Light can be studied, darkness cannot. Darkness cannot be broken down. A simple ray of light tears the darkness and illuminates the surface where the light beam finishes. Dark is a term that we humans have created to describe what happens when there's lack of light."

Finally, the student asked the professor, "Sir, does evil exist?" The professor replied, "Of course it exists, as I mentioned at the beginning, we see violations, crimes and violence anywhere in the world, and those things are evil."

The student responded, "Sir, Evil does not exist. Just as in the previous cases, Evil is a term which man has created to describe the result of the absence of God's presence in the hearts of man."

***

Convergence:

So far, we have gone over the idea of monotheism and spirituality in the world, but not on the direct impact on us. I feel I must expand on the convergence of the intelligence and dominance of humankind, and religion or spirituality's role on that. When asked why we humans are intelligent dominating creatures on this earth, most reply communication, use of tools, actions for pleasure, ability to reason, and religion. These are where I shall expand.

In communication, we are not alone. Ants communicate through antennae, birds through squeaks, snakes through body language, and dolphins with vocal chords, just as we do. In use of tools, we are not alone. Chimpanzees use sticks, and spiders use webs, each to catch food. In actions for pleasure, we are not alone. Dolphins mate for pleasure, and dogs play for pleasure. This leaves religion and ability to reason, two things in which we are alone, yet may be very much the same.

In a way, one could say religion is because of the ability to reason, or that the ability to reason is because of religion. When you reason through right or wrong, perhaps even when you reason through the creation of all things, many often revert to religion. Just as when someone looks into religion, they must be able to reason through it all. It is our need to think abstractly and our need to rationalize everything that we turn to religion to piece it all together, and our reasoning leads us to believe ultimately in religion itself. Some claim that even atheism is somewhat turning into a religion, with rationalizing thought, and organizing of followers.

That being said, I believe it is more rational that the ability to reason and abstract thinking came from religion than the other way around, because religion can play a role in the before mentioned ideas of human's superiority. Religion can communicate thoughts and history, religion can be used as a binding or controlling tool, and religion can be used as an action creating joy. Egyptians for example, used their religion to rationalize the world, control their people, and to record their history- all examples of human intelligence coming out of religion. It almost seems that to be intelligent, to be superior, and to be human is to be religious in one way or another.

The only remaining question is what caused this sensation to begin? I can only imagine the prehistoric human coming across something that they feared yet revered, that human yearning for greater understanding and belief in a higher presence kicked in, and they began worshipping that which they found. Then, as it spread, it encouraged thought and communication while providing comfort and rationality. Scientifically it is the only thing that us from other species and designates us as intelligent. To think is to yearn for belief.

***

Summary:

So now, before I conclude with this first section, we must take a look back at all that has been presented so far. From the laws of physics we have concluded that in order for all the actions and motions in space to be possible, there must have been an initial or constant action from an outside stronger force. From the laws of conservation of mass and energy, something had to have created this expanse of mass, and something must become of our human energy and thoughts that drive us beyond the bounds of our natural capabilities.

From there we looked at the properties of zero and infinity showing the matching properties of both starting with the properties of space, with vast emptiness yet constant expansion, and also the fact that a point in space with zero motion will still move infinitely through the constantly expanding space as well as through time- the forth dimension. Next, we moved to economics, where inflation shows that an infinite number of something would really give it zero value. Mathematics came next showing division and multiplication of both zero and infinity- all coming to the same results.

Next came psychology, where we see that pain leads one to acknowledge happiness, and the world is filled with colors that psychologically make us happy. This showed the unnatural perfections of our behavior and the world to work together to ultimately lead to our happiness. Psychology also showed that over the course of time, and majority in the human race, people have attributed this unnatural perfection to a superior being or God. Then finally, we came to convergence, explaining that religion and the ability to reason are what set us apart from other species in all other ways, and maybe even together.

The evidence to me is unfathomable. Together, all these pieces seem to fit perfectly together to form a puzzle of understanding the reality of a superior force or being. Of course, over time, pieces will be argued for and against, but ultimately the puzzle still points to the same results of the existence of a force or being superior to us, and all the forces of nature. Finally, as in any text about the existence of a God, I must do a basic outline Kurt Godel's Ontological Argument (a mathematical equation that states that on some world or in some universe there is, by necessity a God), which was recently formalized and mathematically "proven" by two computer scientists in Germany. In this model, x represents an object in a given world or universe and P represents an essence of that object, which can represent good or godlike properties. I will list out the equations below, but the meaning of it in the end is that God is that for which no greater can be achieved. Essentially a benchmark. Here it is:

Ax. 1. $\{P(\varphi) \wedge \Box \forall x[\varphi(x) \to \psi(x)]\} \to P(\psi)$

Ax. 2. $P(\neg\varphi) \leftrightarrow \neg P(\varphi)$

Th. 1. $P(\varphi) \to \Diamond \exists x[\varphi(x)]$

Df. 1. $G(x) \iff \forall\varphi[P(\varphi) \to \varphi(x)]$

Ax. 3. $P(G)$

Th. 2. $\Diamond \exists x\, G(x)$

Df. 2. $\varphi \text{ ess } x \iff \varphi(x) \wedge \forall\psi\{\psi(x) \to \Box \forall y[\varphi(y) \to \psi(y)]\}$

Ax. 4. $P(\varphi) \to \Box P(\varphi)$

Th. 3. $G(x) \to G \text{ ess } x$

Df. 3. $E(x) \iff \forall\varphi[\varphi \text{ ess } x \to \Box \exists y\, \varphi(y)]$

Ax. 5. $P(E)$

Th. 4. $\Box \exists x\, G(x)$

-Definition 1: x is God-like if and only if x has as essential properties those and only those properties which are positive
-Definition 2: A is an essence of x if and only if for every property B, x has B necessarily if and only if A entails B
-Definition 3: x necessarily exists if and only if every essence of x is necessarily exemplified
-Axiom 1: Any property entailed by—i.e., strictly implied by—a positive property is positive
-Axiom 2: A property is positive if and only if its negation is not positive
-Axiom 3: The property of being God-like is positive
-Axiom 4: If a property is positive, then it is necessarily positive
-Axiom 5: Necessary existence is a positive property
From these axioms the following theorems are conceived:
-Theorem 1: If a property is positive, then it is consistent, i.e., possibly exemplified.
-Theorem 2: The property of being God-like is consistent.
-Theorem 3: If something is God-like, then the property of being God-like is an essence of that thing.
-Theorem 4: Necessarily, the property of being God-like is exemplified.

Part 2

"Unless we know the value of other religious traditions, it is difficult to develop respect for them. Mutual respect is the foundation of genuine harmony. We should strive for a spirit of harmony, not for political or economic reasons, but rather simply because we realize the value of other traditions. I always make an effort to promote religious harmony." - Dalai Lama

Death, fellowship, purpose in life- these are some of the biggest reasons people seek the knowledge of God or any superior being that tingles inside them. Many of them share common traits that draw together a common thread of what religion really is, and some of them have extraordinary aspects that seem to defy the minds of people. The purpose of this second part is to bring together different views on the same subjects. In short, to bring alternate views on Abrahamic beliefs to different Abrahamic believers, and to bring logical views on Abrahamic traditions to non-Abrahamic believers. Just as the Dalai Lama was saying, it is important to know these views in order to respect, or for that matter disrespect the beliefs of others, while it is also important for those of the Abrahamic faiths to know the common grounds between them and to find reason in all of it. As a disclaimer- this section does go into religion, not just science.

In my personal views, all people of Abrahamic faith worship the same God- the God that Abraham followed, and if someone of Abrahamic faith disagrees, then they are saying there is another God than the one they follow. That would be blasphemy in any Abrahamic faith (since all of these religions say that no other God exists). That being said, all prayers would go to the same God, and that God would hear all prayers- because who else could hear them? It is upon this middle ground that mutual respect, and therefore harmony can be developed.

So this part is about bringing forward information on Abrahamic tradition to both followers and non-followers of those traditions. I have made a great attempt to make it accurate to all Abrahamic faiths including Islam, Christianity, and Judaism. I will start with the beginning of time, and the beginning of tradition in Abrahamic faiths, and work to the more complex issues within these subjects.

***

We begin with the tradition of the universe's creation. According to Abrahamic faiths, the very first thing to happen in this universe was God saying "let there be light," and then there was light, and it just so happens that science would support this theory in the form of acoustic waves. One of the major pieces of evidence for the big bang theory is "cosmic microwave background radiation" picked up through radio telescopes, and it is the sound waves from the big bang theory that cause the galaxies to form in the way they do now. In accordance with that theory, and the religious beliefs, the Abrahamic faiths believe that Hebrew was the language of creation, and is the sacred language of God. Along those same lines, for long periods of history, Hebrew was used only as a sacred language and not used in everyday life, and to this day, many people feel it is sacrilegious to say certain words and phrases in Hebrew, since they would be saying them in

"God's language".

Moving along, if you are aware of computer code, or even the movie "The Matrix" for that matter, it would not be hard to imagine words and code displaying as complex groupings and structures of reality in everything we see around us, but in truth a language itself being this code would be extremely difficult. That being said, Hebrew already is a complex code in its own right, because every word and every letter in the Hebrew language has a numerical value to it. This numerical coding of the language is call Gematria, and while some words of Arabic, Spanish, Greek, and Yiddish can be calculated into Gematria, it is most widespread and easily converted through the Hebrew language.

100 = ק	10 = י	1 = א
200 = ר	20 = כ	2 = ב
300 = שׁ	30 = ל	3 = ג
400 = ת	40 = מ	4 = ד
	50 = נ	5 = ה
	60 = ס	6 = ו
	70 = ע	7 = ז
	80 = פ	8 = ח
	90 = צ	9 = ט

So what we see to this point is a number coded language in which every word, even every sound has a complete numerical value, unlike any other language, and the religious concept of Hebrew being the language of God spoken at creation to form the world and universe. The last of those being a concept from ancient time before computer code or before even a general or scientific concept of the world and universe was developed. Yet still, ancient legend backed by linguistic-numerical code that can be either written, spoken, or counted still creates no basis for the for scientific belief for being the center of all existence and all things… Until you think of the string theory.

The string theory is a theory that was intended to prove together the theories of quantum mechanics and general relativity. It states that at the core of all particles, at the smallest levels of elemental science, are little bands of energy in string like forms. Though it would be a leap of a hypothesis, it could be hypothesized from this theory that these bands of energy at the smallest levels are bands of Hebrew text or linguistic-numerical waves of sounds resembling strings. Each Hebrew string would be energy from audio, or bands of energy containing different properties based on linguistic-numerical value. For example, the Hebrew word "chai" means life, and has the numerical value of 18. So it could be coded into other words through its numerical or linguistic value, just as saying the word "chai" aloud sends audio waves in a unique form, and when spelled out, the Hebrew word of two letters forms a letter string- or word.

Therefore all together, words in Hebrew hold numerical code value, audio sound value, and a physical written value; additionally it is accepted by religious belief as the language of creation. This means it fits every aspect to be possibly the makeup of all existence, the ultimate code to the universe, or the base of all energy. While being a long step to claim it as fact in the theory of string theory, it appears to be a possible contender. It may or may not be worth seeking deeper into this subject, but I feel compelled to make the reasoning available.

***

As we move forward, we go into an idea of extreme controversy. Over the last several decades, the battle of evolution verses creationism has plagued the relationship between religion and science as a two-sided unwavering argument, in which both parties very well may be wrong (I do anticipate just about everyone to taking offense to that last line). This theory of genesis, as I have come to call it, asserts that several religious texts and traditions have suggested what is now known as the theory of evolution, and that evolution and creationism combined can show quite similar attributes when looked at from a deeper level. Put more simply-religious texts somewhat show elements of evolution within the creationist theory. Before I may get into this hybrid creation-evolution theory though, I will explain both the theory of evolution and the theory of creation, so that everyone has a clear understanding.

The theory of evolution, based on natural selection, is quite simple. It states that animals evolved through millions of years due to genetic mutations that helped them in nature, and therefore were passed on to their offspring. For instance, a fish with larger fins could swim away from predators faster, therefore the fish with larger fins didn't get eaten and lived on to reproduce, which led to more fish with larger fins at which time the process repeats itself. There are numerous evidences for this theory including microevolution, which shows that some diseases have mutated or evolved to become immune to some antibiotics, and other evidences such as similar bone structures in animals. For example- our

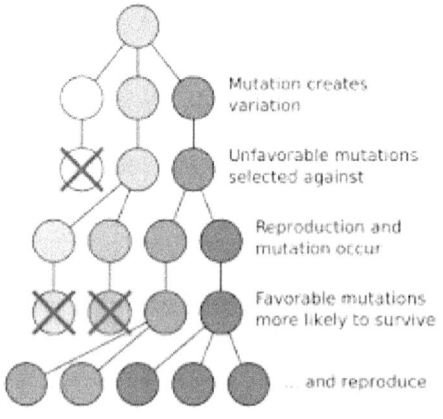

hand, a fishes fin, a birds wing, and an alligators front foot have similar if not identical bone structures, possibly meaning common ancestors. Evidences for this theory go on and on, but the basics of the theory of evolution should be clear.

Creationism on the other hand, mostly provides counter-evidence against evolution asking questions such as, how did life, even single celled organisms just happen to appear or develop without a creator? It also questions the complexity of evolution such as, how did fish change the way the move and the way they breath to become land animals- especially when no land animals have gills or fish tails? (Much like the creation and destruction of energy we talked about earlier, where did this conscience come from?) Then of course, the strongest argument for creationism comes from the believers and the religious texts. The three most widespread and accepted religions in the world have the same story for creation, so that billions of Muslims, billions of Christians, and billions of Jews all agree on their origins on this earth- which obviously includes a creator.

It strikes me as odd that intelligent conscience life could spawn from nothing, or could mutate to become so complex without outside aid. Yet it also strikes me as odd that many creationists refuse to accept scientific evidence of even small mutations and change over time. The problem, I believe, is that so many creationists take the religious texts so literal and word for word, when it is widely known that religious texts are often written in parables and metaphors so they are more easily understood and comprehended. It is widely known that one must look much deeper into the texts to find the meaning of all history. I believe that when broken down, religious texts actually show the first claim of a creationist evolution and through reading and thinking, can prove such a concept. In fact, it has been said that the first theory of natural selection was from a very religious man of Abrahamic belief. A Muslim named Al-Jahiz wrote a book called Kitab al-Hayawan (The Book of Animals) that talked somewhat about natural selection in the 800s AD.

The initial piece of evidence is the serpent, Satan. In the religious texts, God (known as Allah or Yahweh to some) punished the serpent in the Garden of Eden by making all serpents from that day forward crawl on their bellies. Serpents have also been used as an evidence for evolution in that all current serpents have undeveloped legs under their skin that show they once had legs and later, according to that theory, mutated out of them. Thousands of years before Darwin, people did not have the scientific skill to know these under legs were there or what they were. This is the only direct evidence of change occurring to living organisms in the religious texts, but there is evidence of a separate timeline by which different organisms came into existence (meaning Darwin's timeline may have been wrong).

Religious texts claim that God created organisms in the following order: plants, water organisms, birds (or winged animals of the air as it is written), land animals, then finally humans. Scientists generally agree with the water organisms, land animals, and lastly humans potion of the timeline in the theory of evolution, but the other two are where the differences arise. Think for a moment though, about how such a process could work. Photosynthesizing organisms (plants) develop to a point where they can survive while floating or submerged in water, and then the submerged organisms could develop to where they could absorb or eat other organisms, all while not having to breathe the toxins present in the earth's atmosphere at this time like other animals would. This is the point where religious texts become more probable in scientific terms, because fish developing into birds is more likely than animals developing into birds. I say that because fish are not only more likely to jump higher into the air than land animals, but fishes fins more closely resemble wings than land animals arms do, and fishes scales more closely resemble birds feathers than land animals skin does. One can only hypothesize though.

I can already anticipate the counter argument for this theory as being "why would the religious texts not say that there was a change or evolution within these organisms?" from which I will now continue my theory. First is the fact that people in ancient times would have a hard time accepting or understanding such a concept, so it would be more likely to be hidden or coded into the texts for later generations to understand. Then of course is the fact that often, biblical texts speak in metaphors, codes, and parables to get a point across. We know for instance that science shows the world was not created in seven "days," but perhaps these were not literal days. We know days, nights, weeks, and years were not separated until the forth scriptural "day," so we again see this metaphorism and possibility for scientific authentication.

The final question in this theory is, if God created this world, why would he or she change it? That answer comes in two forms. First back to the serpent example from before, which is a direct example of God changing an organism. The second bit of evidence is in the form of Jesus, a prophet of the Muslims and regarded as the messiah to the Christians. In the religious texts Jesus is a carpenter, which symbolizes work never finished. A table built as a table will work its purpose, but a carpenter can always make it stronger, better, and more fitting to its environment. So as the world changed, so must its organisms- an amazing feat of perfect imperfection, for a perfect being.

Therefore, through the Abrahamic religious texts we see that organisms were changed, the order in which they were brought into existence goes along with scientific probability better, the time period in "days" was most likely not literal days, and like carpentry, anything can be changed to be better or better suited for its environment. This is the theory of genesis. The theory that either early religious scholars were the first to theorize evolution, or evolution was supernaturally guided. Darwin himself recognized the inconsistency of chimpanzee's non-development of tools, religion, and art- raising a virtual enigma of how this great intelligence was developed by human beings. Abrahamic faiths seem to unknowingly bridge Darwin's gap. These Abrahamic faiths all talk about the knowledge of good and evil in human existence, suggesting a divine evolution of human thought and actions.

***

The final piece in this chronicle is the story of a man. A man that has caused peace and has been the subject of wars. A man revered by different peoples around the world. For ages, people have bowed to or fought against the idea of this man, yet he remains a key piece of religion and culture as a remarkable being. This man is Jesus of Nazareth. Again, I will leave everything up to your interpretation, giving only different facts and points of view on this man's impact upon us. I will start with the views of those that reject him outright, then go to those who revere him for different reasons.

In doing the research for this section I studied documentaries and books on atheism and some that are not really atheist, but just simply anti-Jesus. Either way the arguments are all the same when aiming to disprove this man. Most are based off of loose interpretations of varying subjects and attempts to connect unrelated dots, often trying to compare Jesus to sun worship. For instance, the comparison of Jesus of Nazareth to different deities throughout time. Not only did these deities exist outside the range and generations of those that followed Jesus, but they also only have a couple of minor similarities- most actually faked. Many atheist try to compare Jesus' life to that of Krishna, Horus, Mithra, and Dionysus by saying that they were born of virgins, crucified, and resurrected, but this is simply not true. All of them have birth stories such as being born from a rock or almond seed, and all of them have death stories, such as killed by an arrow or eaten alive. None of them died as atonement, were born of virgins, nor were they resurrected. The comparisons are mostly just falsified from loose interpretations.

There are also claims that the nativity story was taken from other religions. The fact is that there are things attributed to Jesus that were skewed or guessed by people generations after Jesus existed, often to fit their own convenience. We have no knowledge of whether or not Jesus was born on December 25th, and it is even likely that this is several months off. We also do not know that three kings (or wise men) came and visited him, only that someone visited him with three gifts- an occurrence that actually happened a length of time after his birth (in fact it could be anywhere from two Wise Men to fifty). The date in December and the story of the how many visitors came were added after all of the disciples and apostles were dead, all of which kills the star alignment theory on Jesus promoted by atheists.

Additional theories touted by atheist groups that Christianity is based off of sun worship, are the comparisons between "son" and "sun" and the supposed lack of outside historical records concerning Jesus. Atheists cite the similarities between the fact that Jesus was God's "son" and the Egyptians worshiped the "sun." This is nothing more than a coincidence in the English language. In Greek and Aramaic these similarities to not exist, and trying to find a connection between this is a waste of time. As for records, we turn to the histories written by Flavius Josephus, who was born around the time of Jesus' death. Though Flavius Josephus was neither a Christian nor a Jew his written histories that mention Jesus as an important figure show that Jesus was a widespread figure influencing the area very shortly after his lifetime.

In short, claims of any comparison between Christianity and sun worship are just false. What remains here is Jesus' impact on other religions to show what this man truly meant. We will examine him from many religions points of view from around the world. We will begin with the most obvious place to start, the Jews.

The Jews believed Jesus could not possibly be the messiah because he did not conquer the Romans and did not become king of the holy land. Yet they held a very worldly perspective. While it is true that Jesus never raised an army, never fought the Romans, and never took Judea out of Roman hands, Jesus did do much more to complete the same expectation. Constantine I, born in 272 became the last Emperor to rule from Rome after changing the capital to Constantinople, but did something far greater in becoming the first Roman Emperor to convert to Christianity. Then in 300, Theodosius I took this conversion a step further and made Christianity the official state religion of the Roman Empire. In essence, the entire Roman Empire bowed before Jesus of Nazareth. Therefore, while Jesus never fought the Romans in battle, he still conquered the entire Empire, a much greater feat the Jews could never have expected.

Many of the world's other large religions accept Jesus of Nazareth as a major figure in their religion as well, including Christians, Muslims, and Baha'i. Christianity states that Jesus was the manifestation of God on Earth to be the savior of people's sins and the Messiah the Jewish people had been waiting for. This led to his crucifixion, in which he became a literal sacrifice, the last sacrifice for human sin. Muslims accept Jesus as the Messiah of the Jewish people and the last Jewish prophet before Muhammad. Muslim belief is that the crucifixion was a betrayal by the Jews, and Jesus' body was switched out with that of Judas on the cross. Finally, Baha'i's state that Jesus was in fact the son of God, but was never resurrected after being crucified on the cross. All three religions, which make up a majority of the worlds religious practitioners, accept Jesus of Nazareth in one way or another as the Messiah and an important messenger of God's word.

One of the most fascinating acceptances of Jesus in other religions is that of Buddhism. Many Buddhist teachers, including Zen master Gasan Joseki, have stated that they believe Jesus of Nazareth was an enlightened being, known as a Bodhisattva. In 2001, the Dalai Lama himself, who is revered as a reincarnation of a Bodhisattva himself, stated that he too believed Jesus was an enlightened Bodhisattva. This widespread belief in Buddhism (a religion completely unrelated to the Abrahamic faiths) has even spawned a belief noted by Indira Gandhi, that many Buddhists believe that in the 15 unknown years of Jesus, he may have traveled to India and encountered Buddhists. Either way the acceptance of Jesus as a religious figure into this far eastern religion is significant as it shows the truly remarkable impact of this man upon the world, as the only figure worshiped in this widespread of a manner around the world. This shows a spiritual impact that many have felt from him.

Therefore, as we have seen, the concept of Jesus having Pagan roots based on sun worship and being mentioned only by Christians beyond the 90's AD is simply a fabrication of loosely fitting or completely fabricated facts based on inaccuracies of dates and languages. The Atheist arguments using these concepts have simply chosen to overlook original translations, the work of Flavius Josephus, and other essential facts in order to promote their agenda. Jewish belief that Jesus cannot be messiah because he did not conquer the Romans is up to personal debate over whether or not the Romans converting and following Jesus constitutes conquering or not. In the end, when factoring in Buddhist, Christian, Muslim, and Baha'I beliefs it becomes apparent that this man had a widespread physical and spiritual impact upon the world. How we interpret that impact or choose to proceed with that is completely up to us as individuals and as peoples of different faiths. I'm sure the debate will continue for centuries, as we will know all the facts of this world, but it is up to us to decide our personal beliefs on this remarkable man.

Closing:

"A man is what he believes" - Anton Chekhov

In closing, I release you to make your criticisms, and to take or leave what you wish. The final thought I will leave all of you with is to leave your heart and mind open to realize what it knows is real- not what I or any other tells you. Whether you want to believe what it is telling you or not, the more you resist it, the more you will find yourself unhappy, arguing with your own conscience. I have lain out what I believe to be real and it is now time that you choose for yourself what the reality of this world we live in is. Around the world, generations of people have chosen a reality, but it is up to you whether you choose to accept this, and the more we turn away from this, the more we are destroying this cultural feature for future generations.

At this point, I hope you, as the reader, have begun to question or think about everything you read in here. While I would ask that you keep an open mind on it all, I know that this is truly the point in which it is best to question it. To me, though this work is incomplete, I have chosen my reality. I believe in an all-powerful God, and I know he or she loves me. I am at ease with my conscience. Now it is your turn.

I do hope that in some small way, this writing helps you or someone else to help shape what you or they believe or disbelieve, and while it would be nice to know that some of you believe in what I believe, it is your choice. You have to be open to accept even that which seems painful at times to accept. Let your heart let go of grudges that limit your thinking. Let your soul tell you what is true.

www.ingramcontent.com/pod-product-compliance
Lightning Source LLC
Chambersburg PA
CBHW070926180526
45168CB00005B/2163